With a few common materials you can provide 37 delightful board games for your first grade classroom. These games are designed to provide important practice for children learning their math facts and algorithms. Hopefully these games will help children memorize their facts in an enjoyable, yet effective, way and be substituted for some of the skill practice sheets that are often provided. The games are designed to be easy to learn, with directions provided on each game board—a perfect choice for learning centers when students must work independently. If you are searching for ways to involve parents or guardians in their children's math experiences, create "math take-home game kits" by mounting the game boards on file folders and attaching small envelopes for game pieces. The games can be used as "homework" assignments—to be played with an adult or older child at home. They can also be sent home for extra practice when parents request helpful materials or when you feel it is needed. Children may even choose to play one of these games at recess or during free time. If you are searching for curriculum materials for after-school programs, summer school programs, or neighborhood center programs, these game materials require minimal preparation.

Before distributing the game materials to students, read the game directions ahead of time and prepare the materials as needed. Most of the games use "cards" that are supplied at the end of *Math Practice Games Grade 1* and indicated in the game directions. To prepare the game cards, photocopy the pattern pages, then cut apart the cards. As you prepare the game boards and cards, be sure to consider these possibilities:
- Copy the game board and cards on construction paper so they will last longer. Also it is not possible to read the problem through the back of the card.
- Enlarge the game board and/or cards if the school copier uses 11" x 17" (279 x 432 mm) paper.
- Laminate the game materials for durability.

Most games also need game markers for the children to use. The game markers must fit in the spaces as indicated on the game board. They must also be easy for children to pick up. Different colors or kinds must be available so the players can distinguish their game markers from their opponents. It is possible to use crayons instead of game markers on many game boards, however, by using game markers the cost for producing the games is minimal and less paper is used in the long run. Here are some items that could be used as game markers:
- counters
- buttons
- macaroni
- paper clips
- candy pieces
- dried beans
- centimeter cubes
- wads or squares of construction paper
- Styrofoam packing materials
- game markers from commercial games

Undoubtedly, you will be able to add to this list!

Math Practice Games Grade 1 can be used in many different ways. If a game is being introduced to the whole class, reproduce the game board on a transparency and use the overhead projector when explaining the game rules. Some teachers may prefer to introduce a game to a math group and have just that group use it. At times, a game may simply be given to a pair or group of children who can read the directions and follow them on their own. If a child cannot read the directions, he or she can be paired with someone who can.

When children are playing a game, they should be expected to check each other's answers and to use scratch paper if necessary. Figuring out fact answers mentally or memorizing them is the goal at some point. The game cards can be used as flash cards or even as timed tests for straightforward practice of memorization.

Hopefully the materials in *Math Practice Games Grade 1* will become a useful part of your overall math program and provide delightful math experiences for children who have not yet mastered their facts or computation skills.

FOUR TOGETHER

Players: 2

Object: To place 4 game markers together

Other Materials: Game markers or different color crayon for each player, Cards A and B

To Play:
1. Mix up the cards and place them *face down* in a pile.
2. Each player takes a card, calls out the answer, and marks one of the squares with the correct answer.
3. If no answer is left, the player loses that turn.
4. Continue playing until someone marks 4 squares together.

9	6	1	8	3	4	7	8	
5								2
4								5
3								8
10								6
6								4
2								2
0								5
7								10
1								4
5	9	7	6	3	7	9	1	

GET FOUR FACTS

Players: 2
Object: To place 4 cards on top of a number
Other Materials: Cards A and B

To Play:
1. Place the cards *face down* and mix them up.
2. Give each player a copy of the game board.
3. Each player takes a card and places it on the answer on his or her game board.
4. If the answer is not on the grid, the player loses that turn.
5. Keep on playing until someone has 4 cards on top of a number.

7	6	10
5	9	8

Train from Scaryville

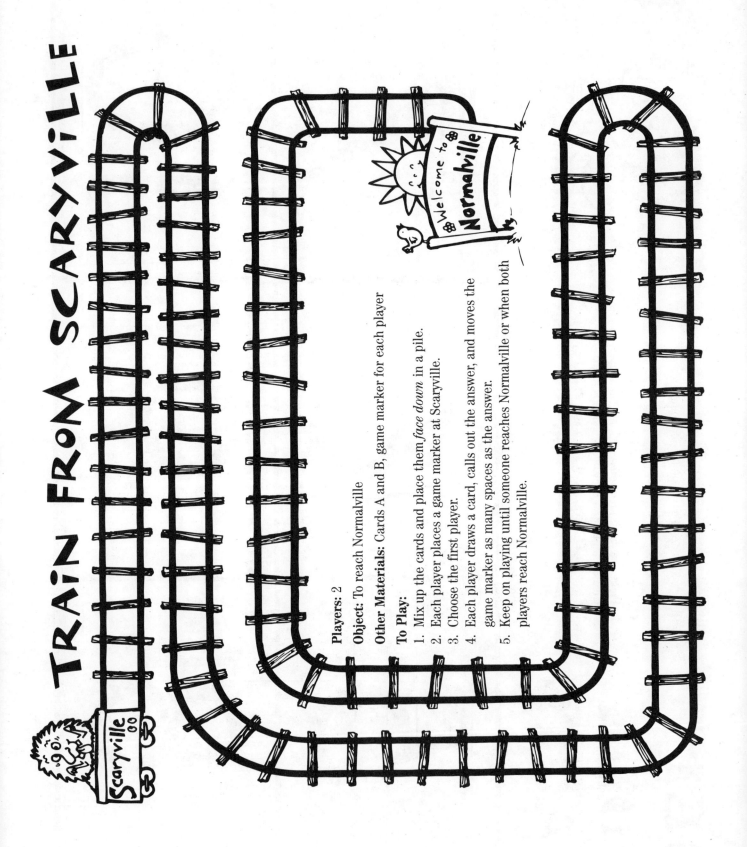

Players: 2

Object: To reach Normalville

Other Materials: Cards A and B, game marker for each player

To Play:
1. Mix up the cards and place them *face down* in a pile.
2. Each player places a game marker at Scaryville.
3. Choose the first player.
4. Each player draws a card, calls out the answer, and moves the game marker as many spaces as the answer.
5. Keep on playing until someone reaches Normalville or when both players reach Normalville.

TWO PLAYER BINGO

Players: 2
Object: The first player to have 4 markers in a row is the winner.
Other Materials: Cards A and B, game markers

To Play:
1. Mix up the cards and place them *face down* in a pile.
2. Each player chooses a bingo board.
3. Each player draws a card, calls out the answer, and marks one of the squares with the correct answer on his or her bingo board.
4. If no answer is left, the player loses that turn.
5. Keep on playing until someone covers 4 squares in a →, ↓ or ↗ row.
6. Play again. Keep track of who wins each time.

10	5	3	8
1	8	7	4
7	6	10	9
9	6	2	5

Name _____

9	6	8	3
10	2	7	9
7	6	1	5
4	8	10	5

Name _____

© Instructional Fair • TS Denison   IF5203 *Math Practice Games Gr. 1*

BEAR BINGO

Players: Entire class

Object: The first player to cover 4 squares in a row is the winner.

Other Materials: Cards A and B, game markers

Prepare: Give each player a copy of the bingo board.

To Play:
1. Mix up the cards and place them *face down* in a pile.
2. The teacher draws a card and calls out the problem or writes it on the chalkboard.
3. Each player marks one of the squares with the correct answer on his or her bingo board.
4. Keep on playing until someone covers 4 squares in a →, ↓ or ↗ row.

9	8	10	4	3	10
4	5	7	8	6	9
5	2	10	1	7	6
9	8	3	6	2	7
10	8	9	9	10	8
4	5	6	7	0	10

GET A TOUCHDOWN

Players: 2 or more
Object: To reach "Touchdown!"
Other Materials: Cards C, game marker for each player

To Play:
1. Place the cards *face down* and mix them up.
2. Each player places a game marker at "Start."
3. Each player draws a card, calls out the answer, and moves his or her marker to the first answer.
4. If that number is already covered, the player loses that turn.
5. Keep on taking turns until someone gets to "Touchdown!"

GET FOUR FACTS AGAIN

Players: 2 or more
Object: To place 4 cards on top of a number
Other Materials: Cards C (without "7 – 0")

To Play:
1. Place the cards *face down* and mix them up.
2. Give each player a copy of the game board.
3. Each player draws a card, calls out the answer, and places the card on that number on her or his game board.
4. Keep on playing until someone places 4 cards on top of a number.

0	1	2
3	4	5 and 6

TWO PLAYER "X" BINGO

Players: 2
Object: The first player to have 4 markers in a row is the winner.
Other Materials: Cards C, game markers

To Play:
1. Mix up the cards and place them *face down* in a pile.
2. Each player chooses a bingo board.
3. Each player draws a card, calls out the answer, and marks one of the squares with the correct answer on his or her bingo board.
4. If no answer is left, the player loses that turn.
5. Continue playing until someone has 4 markers in a →, ↓, or ↗ row.
6. Play again. Keep track of who wins each time.

© Instructional Fair • TS Denison

TRUCK BINGO

Players: Entire class
Object: The first player to mark 4 squares in a row is the winner.
Other Materials: Cards C, game markers
Prepare: Give each player a copy of the bingo board.
To Play:
1. Mix up the cards and place them *face down* in a pile.
2. The teacher draws a card and calls out the problem or writes it on the chalkboard.
3. Each player marks one of the squares with the correct answer on his or her bingo board.
4. Continue playing until someone covers 4 squares in a →, ↓ or ↗ row.

3	0	3	6	4	1
1	5	2	7	0	2
0	0	1	0	2	1
4	3	6	2	0	4
5	0	1	1	4	3
3	2	5	0	2	1

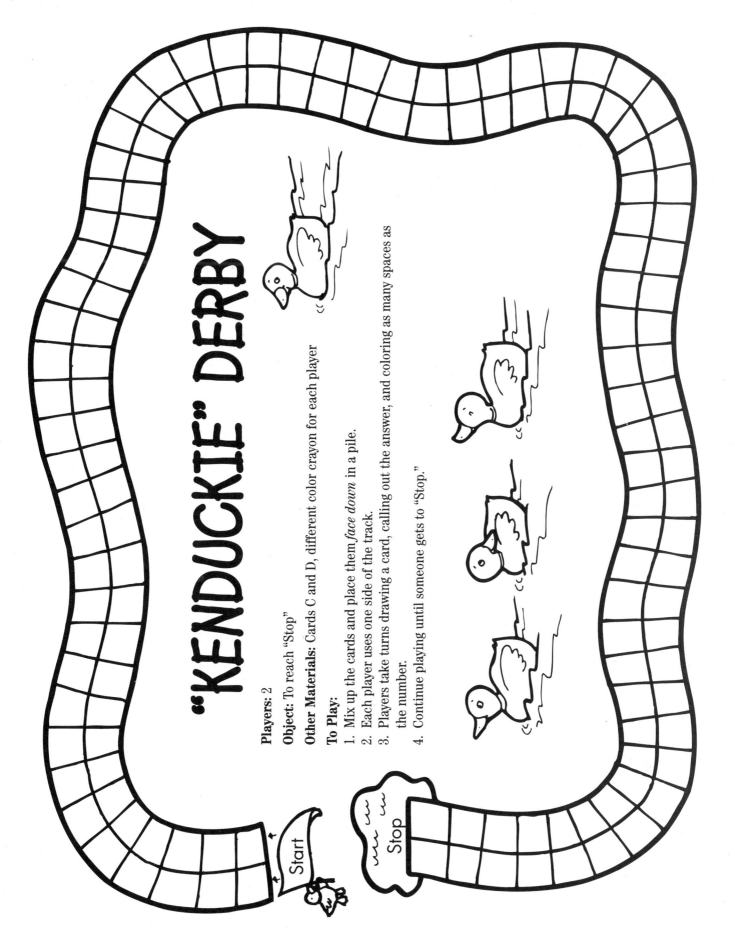

TAKE OFF!

Players: 2
Object: To mark 4 spaces on a rocket
Other Materials: Cards C and D, game markers or different color crayons
To Play:
1. Mix up the cards and place them *face down* in a pile.
2. Each player draws a card, gives the answer, and marks that answer on a rocket.
3. If no answer is left, the player loses that turn.
4. Continue playing until someone marks 4 spaces on a rocket for "take off."

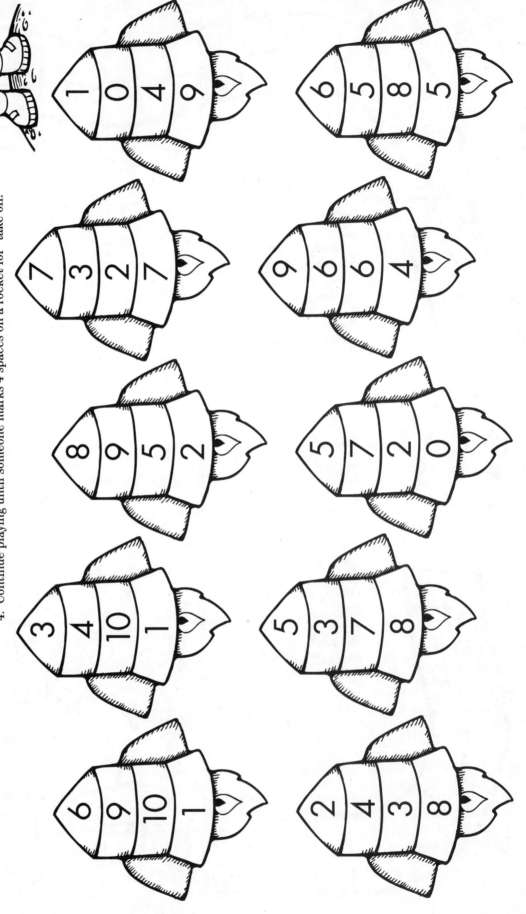

Circle Bingo

Players: 2

Object: The first player to place 4 markers in a row is the winner.

Other Materials: Cards C and D (without "10 – 0"), game markers

To Play:
1. Mix up the cards and place them *face down* in a pile.
2. Each player chooses a bingo board.
3. Each player draws a card, calls out the answer, and marks one of the circles with the correct answer on her or his bingo board.
4. If no answer is left, the player loses that turn.
5. Take turns until someone covers 4 circles in →, ↓ or ↗ row.
6. Play again. Players keep track of who wins each time.

Name _____

2	1	4	2
5	9	0	7
0	3	8	4
5	1	3	6

Name _____

7	4	5	0
3	2	1	6
0	8	4	2
9	1	5	3

CAKE BINGO

Players: Entire class

Object: The first player to cover 4 squares in a row is the winner.

Other Materials: Cards C and D, game markers

Prepare: Give each player a copy of the bingo board.

To Play:
1. Mix up the cards and place them *face down* in a pile.
2. The teacher draws a card and calls out the problem or writes it on the chalkboard.
3. Each player marks one of the squares with the correct answer on her or his bingo board.
4. Continue playing until someone covers 4 squares in a →, ↓ or ↗ row.

SHOOTING BASKETS

Players: 2

Object: The first player with the most points is the winner.

Other Materials: Cards E, different color crayon or game markers for each player

To Play:
1. Mix up the cards and place them *face down* in a pile.
2. Each player draws a card, calls out the answer, and marks one of the spaces with the correct number.
3. If no answer is left, the player loses that turn.
4. When a player marks all 3 spaces on a basket, the player gets 2 points.
5. Play continues until all spaces are marked.
6. Each player adds up his or her points.

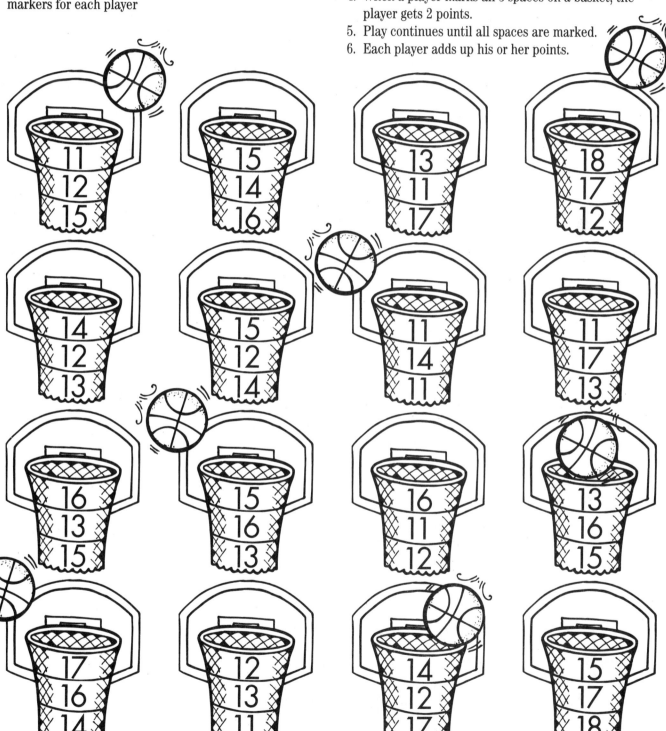

© Instructional Fair • TS Denison

"DINOPLUSICUS"

Players: 2
Object: To cover 4 teeth in a row
Other Materials: Cards E, different color crayon or game markers for each player
To Play:
1. Mix up the cards and place them *face down* in a pile.
2. Each player draws a card, calls out the sum, and marks one tooth with the correct answer.
3. If no answer is left, the player loses that turn.
4. Continue taking turns until someone has marked 4 teeth together.

DIAMOND BINGO

Players: 2
Object: To cover 4 diamonds in a row
Other Materials: Cards E, game markers for each player

To Play:
1. Mix up the cards and place them *face down* in a pile.
2. Each player draws a card, calls out the sum, and marks the correct answer on his or her bingo board.
3. If no answer is left, the player loses that turn.
4. Continue taking turns until someone covers 4 diamonds in a →, ↓, or ↗ row.
5. Play again. Keep track of who wins each time.

BIGFOOT BINGO

Players: Entire class

Object: The first player to mark 4 squares in a row is the winner.

Other Materials: Cards E, game markers

Prepare: Give each player a copy of the bingo board.

To Play:
1. Mix up the cards and place them *face down* in a pile.
2. The teacher draws a card and calls out the problem or writes it on the chalkboard.
3. Each player marks one of the squares with the correct answer on his or her bingo board.
4. Continue playing until someone covers 4 squares in a →, ↓ or ↗ row.

12	15	14	13	14	11
11	13	11	15	16	12
12	14	18	12	11	13
13	11	13	17	14	12
16	12	11	15	11	17
14	11	16	12	13	15

HAUNTED HOUSE VISIT

Players: 2

Object: The first player to mark all 7 places in the house is the winner.

Other Materials: Cards A and E or B and E, game markers for each player

To Play:
1. Mix up the cards and place them *face down* in a pile.
2. Each player draws a card, calls out the sum, and marks the place with the correct answer.
3. Keep on taking turns until someone marks all 7 places in the house.

COUNT TWO MORE

Players: 2

Object: To mark one whole "2"

Other Materials: Cards G, game markers or different color crayon for each player

To Play:
1. Mix up the cards and place them *face down* in a pile.
2. Each player draws a card and marks one of the spaces with the number that is 2 more than the number on the card.
3. If no answer is left, the player loses that turn.
4. Keep on playing until someone marks all spaces on a "2" with his or her game markers.

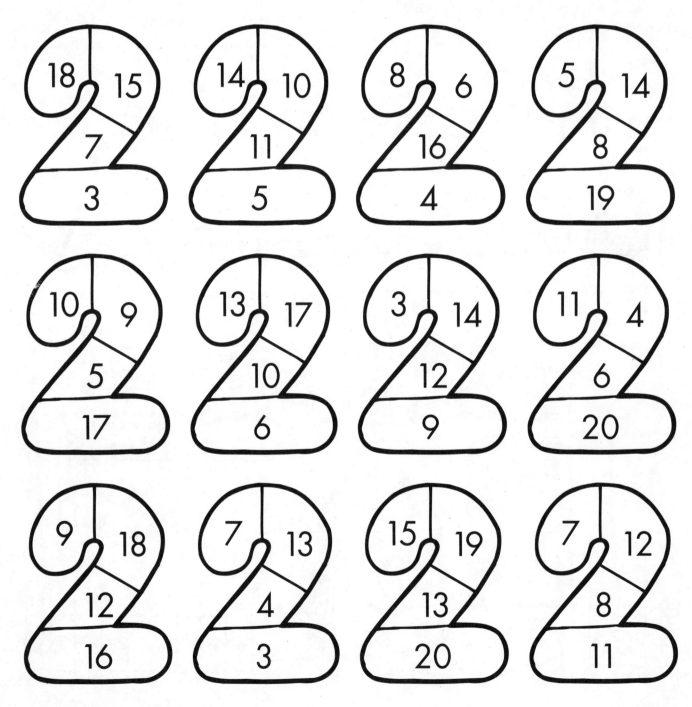

FAT CAT BINGO

Players: Entire class

Object: The first player to cover 4 spaces is the winner.

Other Materials: Cards A and E or B and E, game markers

Prepare: Give each player a copy of the bingo board.

To Play:
1. Mix up the cards and place them *face down* in a pile.
2. The teacher draws a card and calls out the problem or writes it on the chalkboard.
3. Each player marks one of the squares with the correct answer on his or her bingo board.
4. Keep on playing until someone marks 4 squares in a →, ↓ or ↗ row.

10	12	3	7	8	16
17	6	12	10	2	13
4	7	11	15	9	6
16	4	1	17	14	3
10	15	8	5	0	11
5	2	9	13	14	18

VISIT THE SOLAR SYSTEM!

Players: 2
Object: To visit all 10 parts of the solar system
Other Materials: Cards F, game markers for each player

To Play:
1. Place the cards *face down* and mix them up.
2. Each player takes a card, calls out the answer, and marks the place with the correct number on his or her game board.
3. If no answer is left, the player loses that turn.
4. Keep on playing until someone marks all 10 parts of the solar system.

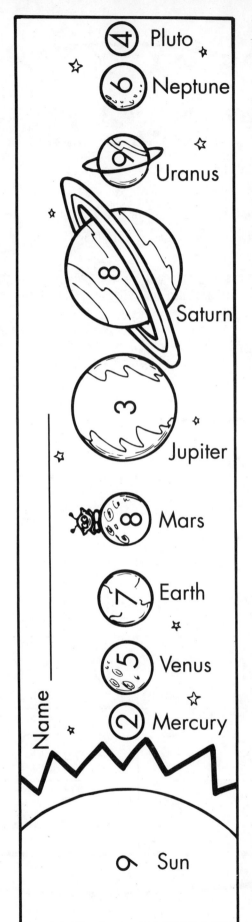
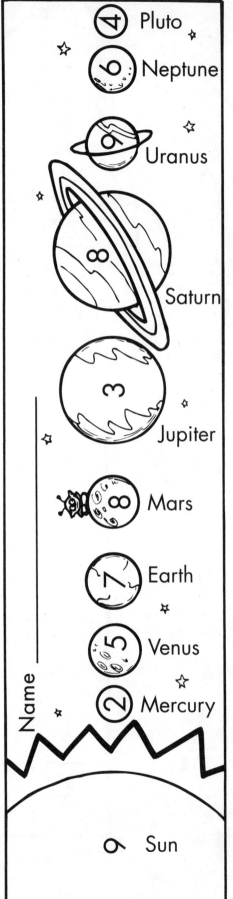

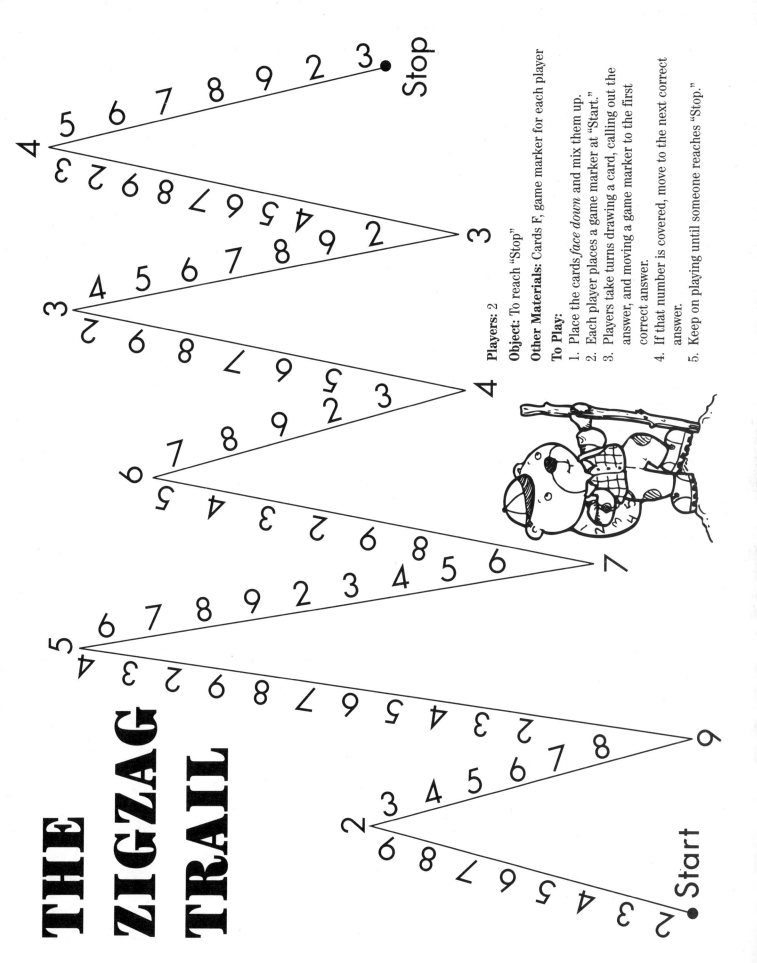

APPLE BINGO

Players: 2
Object: To cover 4 apples in a row
Other Materials: Cards F, game markers for each player

To Play:
1. Mix up the cards and place them *face down* in a pile.
2. Each player chooses a bingo board.
3. Each player takes a card, calls out the answer, and marks one of the apples with the correct number on his or her bingo board.
4. If no answer is left, the player loses that turn.
5. Keep on playing until someone covers 4 apples in a →, ↓ or ↗ row.
6. Play again. Keep track of who wins each time.

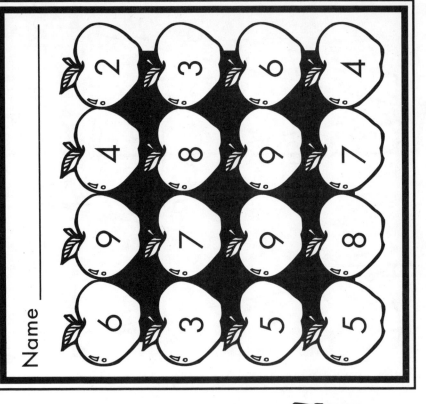

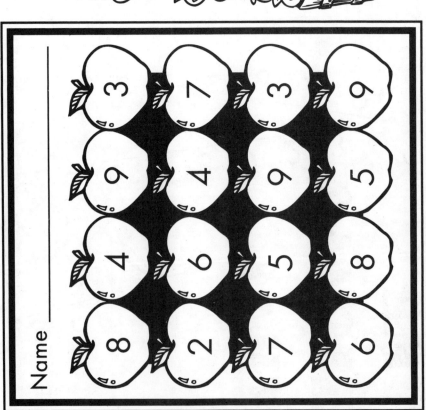

EAT A PIZZA!

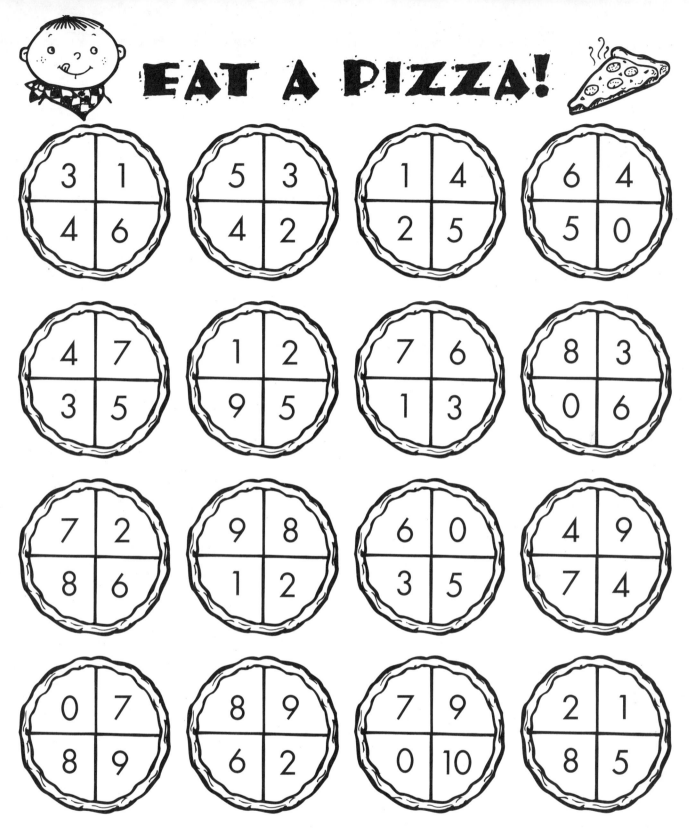

Players: 2

Object: To mark a whole pizza

Other Materials: Game markers or different color crayon for each player, Cards C, D, and F

To Play:
1. Mix up the cards and place them *face down* in a pile.
2. Each player draws a card, calls out the answer, and marks one of the pizza slices with the correct number.
3. If no answer is left, the player loses that turn.
4. Keep on playing until someone marks a whole pizza.

COLLECT THE FACTS

Name _____

0	5
1	6
2	7
3	8
4	9 and 10

Players: 2
Object: The first player to cover all spaces on a grid is the winner.
Other Materials: Game markers, Cards C, D, and F

To Play:
1. Mix up the cards and place them *face down* in a pile.
2. Each player draws a card, calls out the answer, and places it on the correct number on her or his grid.
3. If no answer is left, the player loses that turn.
4. Keep on playing until someone covers all the answers on a grid.
5. Play again. Keep track of who wins each time.

Name _____

0	5
1	6
2	7
3	8
4	9 and 10

ELEPHANT BINGO

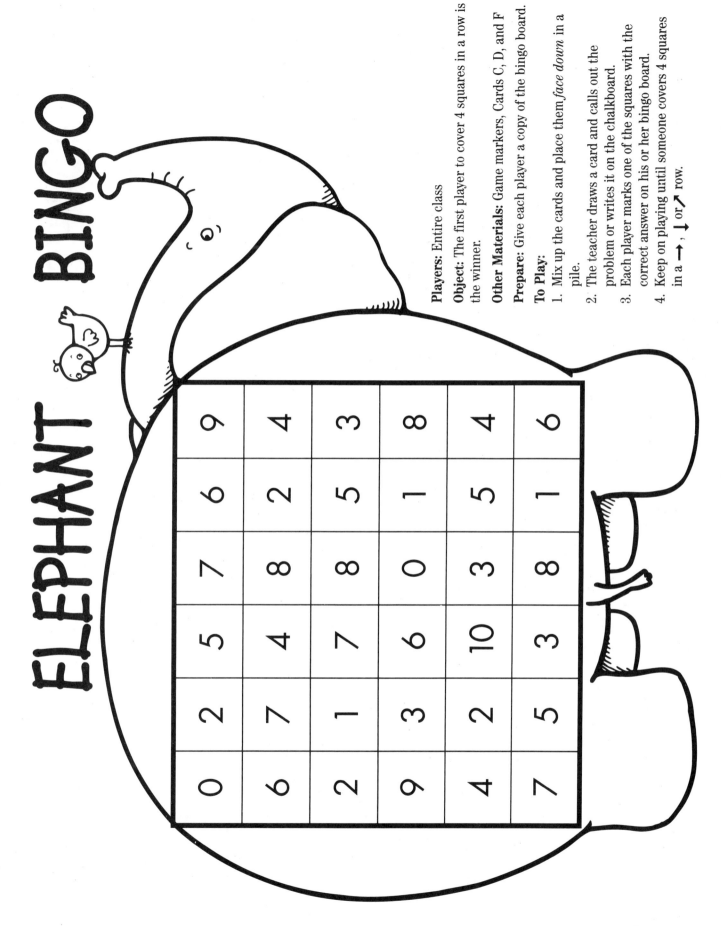

Players: Entire class
Object: The first player to cover 4 squares in a row is the winner.
Other Materials: Game markers, Cards C, D, and F
Prepare: Give each player a copy of the bingo board.
To Play:
1. Mix up the cards and place them *face down* in a pile.
2. The teacher draws a card and calls out the problem or writes it on the chalkboard.
3. Each player marks one of the squares with the correct answer on his or her bingo board.
4. Keep on playing until someone covers 4 squares in a →, ↓ or ↗ row.

0	2	5	7	6	9
6	7	4	8	2	4
2	1	7	8	5	3
9	3	6	0	1	8
4	2	10	3	5	4
7	5	3	8	1	6

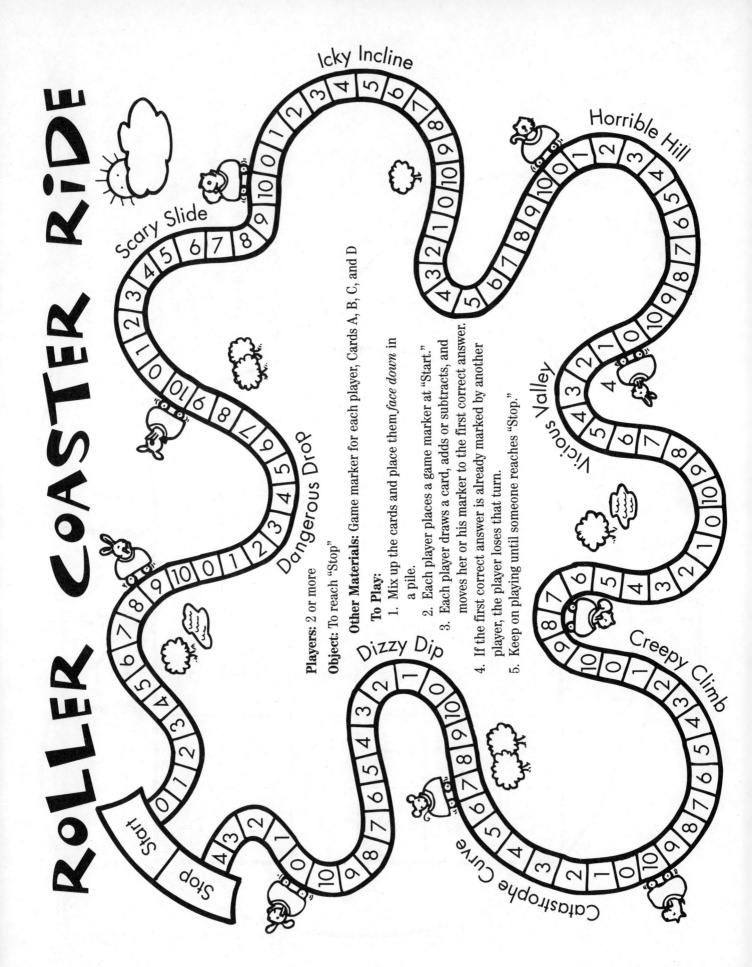

STAR BINGO

Players: 2
Object: To cover 4 stars
Other Materials: Game markers, Cards A, B, C, and D

To Play:
1. Mix up the cards and place them *face down* in a pile.
2. Each player draws a card, calls out the answer, and marks one of the stars with the correct number on his or her bingo board.
3. If no answer is left, the player loses that turn.
4. Keep on playing until someone covers 4 stars in a →, ↓ or ↗ row.
5. Play again. Keep track of who wins each time.

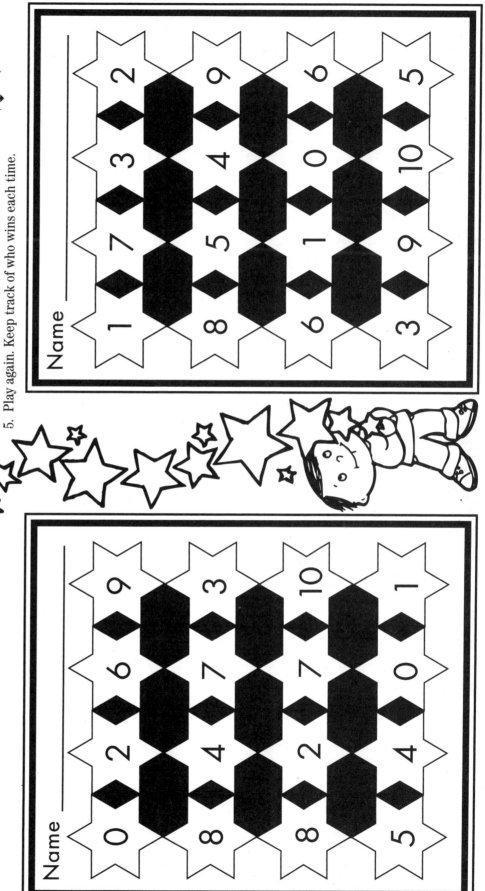

TOUCAN PLAY

Toucan 1 (top left): 18, 9, 3, 5
Toucan 2 (top right): 4, 8, 15, 2
Toucan 3 (middle left): 13, 8, 4, 16
Toucan 4 (middle center): 12, 10, 7, 1
Toucan 5 (middle right): 10, 6, 7, 14
Toucan 6 (bottom left): 9, 12, 5, 17
Toucan 7 (bottom center): 14, 11, 2, 0
Toucan 8 (bottom right): 11, 13, 6, 3

Players: 2
Object: To mark a whole toucan bill
Other Materials: Game markers or different color crayon for each player, Cards A, B, C, D, E, and F

To Play:
1. Mix up the cards and place them *face down* in a pile.
2. Each player draws a card, calls out the answer, and marks one space with the correct number.
3. If no answer is left, the player loses that turn.
4. Continue playing until someone covers all spaces on a toucan's bill.

© Instructional Fair • TS Denison

IF5203 *Math Practice Games Gr. 1*

CLIMB MATH MOUNTAIN

Players: 2 or more

Object: To reach the top of the mountain

Other Materials: Game marker for each player, Cards A, B, C, D, E, and F

To Play:
1. Mix up the cards and place them *face down* in a pile.
2. Each player places a game marker at "Start."
3. Each player draws a card and calls out the answer. The player moves to the next level and places her or his marker on the correct answer.
4. If no answer is available on that part of the mountain, the player loses that turn to move a game marker.
5. Continue playing until someone reaches the top of the mountain.

CRAZY QUILT BINGO

Players: Entire class

Object: The first player to cover 4 squares in a row is the winner.

Other Materials: Game markers, Cards A, B, C, D, E, and F

Prepare: Give each player a copy of the bingo board.

To Play:
1. Mix up the cards and place them *face down* in a pile.
2. The teacher draws a card and calls out the problem or writes it on the chalkboard.
3. Each player marks one of the squares with correct answer on his or her bingo board.
4. Keep on playing until someone covers 4 squares in a →, ↓ or ↗ row.

TIC-TAC-TALLY

Players: 2

Object: To place 3 markers in a row

Other Materials: Game markers for each player, Cards H

To Play:
1. Mix up the cards and place them *face down* in a pile.
2. Each player draws a card, calls out the answer, and marks one of the spaces with that number.
3. Keep on playing until someone covers 3 numbers in a →, ↓ or ↗ row.
4. Play again. Keep track of who wins each time.

2	8	5
12	2	17
13	6	9

7	9	15
16	1	5
11	14	10

6	18	11
4	14	7
18	13	3

16	15	4
17	10	1
12	8	3

WIGGLE WORMS

Players: 2

Object: Be the first player to mark 3 spaces on worms. The player with the most points is the winner.

Other Materials: Game markers or different color crayon for each player, Cards H

To Play:
1. Place the cards *face down* and mix them up.
2. Each player takes a card, calls out the answer, and marks one of the spaces with the correct answer.
3. Continue playing until all cards have been used.
4. A player earns a point by marking 3 spaces on a worm.
5. When the game ends, the players count their points. Whoever has more points is the winner.

FIVE IN A ROW

Players: 2

Object: The first player to show 5 numbers in order on her or his grid is the winner.

Other Materials: Cards I

To Play:
1. Mix up the cards and place them *face down* in a pile.
2. Each player takes a card and places it on his or her board in order.
3. If that number does not belong on the board, the player loses that turn and returns the card to the draw pile.
4. Continue playing until someone shows 5 numbers in order on the board.
5. Play again. Keep track of who wins each time.

Name _____   Name _____

9	19	29	14	24	34

NUMBER COVER-UP

Players: 2

Object: The first player to cover all spaces on her or his grid is the winner.

Other Materials: Cards I and J

To Play:
1. Place the cards *face down* and mix them up.
2. Each player takes a card, reads the number, and places the card on the corresponding space on her or his grid.
3. If no answer is left, the player loses that turn.
4. Keep on playing until someone covers all spaces on a grid.
5. Play again. Keep track of who wins each time.

Name _____

Less than 8	Less than 18
Less than 44	Less than 63
Greater than 10	Greater than 26
Greater than 48	Greater than 60

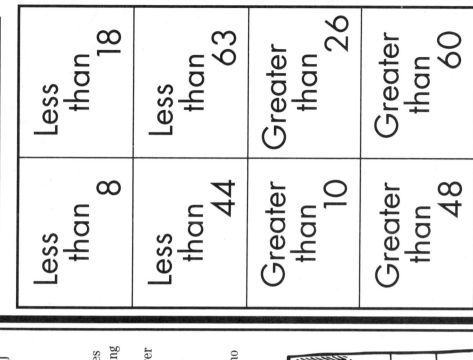

Name _____

Less than 8	Less than 18
Less than 44	Less than 63
Greater than 10	Greater than 26
Greater than 48	Greater than 60

CLIMB THE EMPIRE STATE BUILDING!

	Winner	
	12	
	11	
	10	
	9	
	8	
	7	
	6	
	5	
	4	
	3	
	2	
	1	
	Start	

Players: 2

Object: To reach the "Winner" floor

Other Materials: Cards I and J, game marker for each player

To Play:
1. Mix up the cards and place them *face down* in a pile.
2. Each player places a game marker at "Start."
3. For each round of play, each player takes a card and reads the number.
4. The player with the higher number moves his/her game marker up a floor.
5. If the numbers are equal, discard the cards and draw new cards.
6. Keep on playing until someone reaches the "Winner" floor.

✂ Addition Facts 0–10 ✂ Cards A

5 + 0	1 + 1	7 + 1	5 + 2	5 + 3	5 + 5
4 + 0	10 + 0	6 + 1	4 + 2	4 + 3	6 + 4
3 + 0	9 + 0	5 + 1	3 + 2	3 + 3	5 + 4
2 + 0	8 + 0	4 + 1	2 + 2	8 + 2	4 + 4
1 + 0	7 + 0	3 + 1	9 + 1	7 + 2	7 + 3
0 + 0	6 + 0	2 + 1	8 + 1	6 + 2	6 + 3

Addition Facts 0–10 Reversed — Cards B

0 + 0	0 + 6	1 + 2	1 + 8	2 + 6	3 + 6
0 + 1	0 + 7	1 + 3	1 + 9	2 + 7	3 + 7
0 + 2	0 + 8	1 + 4	2 + 2	2 + 8	4 + 4
0 + 3	0 + 9	1 + 5	2 + 3	3 + 3	4 + 5
0 + 4	0 + 10	1 + 6	2 + 4	3 + 4	4 + 6
0 + 5	1 + 1	1 + 7	2 + 5	3 + 5	5 + 5

Subtraction Facts 0–7 Cards C

2 – 2	4 – 1	5 – 2	6 – 2	7 – 1	7 – 7
2 – 1	4 – 0	5 – 1	6 – 1	7 – 0	7 – 6
2 – 0	3 – 3	5 – 0	6 – 0	6 – 6	7 – 5
1 – 1	3 – 2	4 – 4	5 – 5	6 – 5	7 – 4
1 – 0	3 – 1	4 – 3	5 – 4	6 – 4	7 – 3
0 – 0	3 – 0	4 – 2	5 – 3	6 – 3	7 – 2

Subtraction Facts 1-10 Cards D

8 − 5	9 − 2	9 − 8	10 − 4	10 − 10	10 − 4
8 − 4	9 − 1	9 − 7	10 − 3	10 − 9	10 − 5
8 − 3	9 − 0	9 − 6	10 − 2	10 − 8	10 − 6
8 − 2	8 − 8	9 − 5	10 − 1	10 − 7	10 − 7
8 − 1	8 − 7	9 − 4	10 − 0	10 − 6	10 − 8
8 − 0	8 − 6	9 − 3	9 − 9	10 − 5	10 − 9

Addition Facts 11–18 — Cards E

4 + 7	6 + 6	7 + 6	7 + 7	6 + 9	9 + 9
5 + 6	7 + 5	8 + 5	8 + 6	7 + 8	8 + 9
6 + 5	8 + 4	9 + 4	9 + 5	8 + 7	9 + 8
7 + 4	9 + 3	3 + 9	4 + 9	9 + 6	7 + 9
8 + 3	2 + 9	4 + 8	5 + 8	5 + 9	8 + 8
9 + 2	3 + 8	5 + 7	6 + 7	6 + 8	9 + 7

Subtraction Facts 11–18 — Cards F

11 – 4	12 – 6	13 – 7	14 – 7	15 – 6	18 – 9
11 – 5	12 – 7	13 – 8	14 – 8	15 – 7	17 – 8
11 – 6	12 – 8	13 – 9	14 – 9	15 – 8	17 – 9
11 – 7	12 – 9	12 – 3	13 – 4	15 – 9	16 – 7
11 – 8	11 – 2	12 – 4	13 – 5	14 – 5	16 – 8
11 – 9	11 – 3	12 – 5	13 – 6	14 – 6	16 – 9

✂ Numbers 1–18 ✂ Cards G

1	2	3	4	5	6
7	8	9	10	11	12
13	14	15	16	17	18
1	2	3	4	5	6
7	8	9	10	11	12
13	14	15	16	17	18

✂ Numbers 1–18 ✂

Number Words and Tallies 1–18 Cards H

six	twelve	eighteen	ꟾꟾꟾꟾ ꟾ	ꟾꟾꟾꟾ ꟾꟾꟾꟾ ꟾꟾ	ꟾꟾꟾꟾ ꟾꟾꟾꟾ ꟾꟾꟾꟾ ꟾꟾꟾ
five	eleven	seventeen	ꟾꟾꟾꟾ	ꟾꟾꟾꟾ ꟾꟾꟾꟾ ꟾ	ꟾꟾꟾꟾ ꟾꟾꟾꟾ ꟾꟾꟾꟾ ꟾꟾ
four	ten	sixteen	ꟾꟾꟾꟾ	ꟾꟾꟾꟾ ꟾꟾꟾꟾ	ꟾꟾꟾꟾ ꟾꟾꟾꟾ ꟾꟾꟾꟾ ꟾ
three	nine	fifteen	ꟾꟾꟾ	ꟾꟾꟾꟾ ꟾꟾꟾꟾ	ꟾꟾꟾꟾ ꟾꟾꟾꟾ ꟾꟾꟾꟾ
two	eight	fourteen	ꟾꟾ	ꟾꟾꟾꟾ ꟾꟾꟾ	ꟾꟾꟾꟾ ꟾꟾꟾꟾ ꟾꟾꟾꟾ
one	seven	thirteen	ꟾ	ꟾꟾꟾꟾ ꟾꟾ	ꟾꟾꟾꟾ ꟾꟾꟾꟾ ꟾꟾꟾ

✂ Numbers 1–36 ✂ Cards 1

6	12	18	24	30	36
5	11	17	23	29	35
4	10	16	22	28	34
3	9	15	21	27	33
2	8	14	20	26	32
1	7	13	19	25	31

✂ Numbers 37-72 ✂   Cards J

42	48	54	60	66	72
41	47	53	59	65	71
40	46	52	58	64	70
39	45	51	57	63	69
38	44	50	56	62	68
37	43	49	55	61	67